The Micro Measurement System

A proposal for ending uncertainty in scientific metrics 2014

JOSIAH JAMES INGALLS

ISBN-10: 061595779X
ISBN-13: 978-0-615-95779-1

DEDICATION

This work is dedicated to Albert Einstein, who unified space and time, and began the work of unifying the world of metrology. I'm not yet done with this work, but I hope to eventually finish the masters were.

TABLE OF CONTENTS

1 PREFACE

In measuring extremely small dimensions from the atomic level downward to quanta, science has infinitesimally divided the meter into nanometers, picometers, femtometers and beyond. However, there is a fundamental problem in doing this in that the meter itself is an inexact measure. Hence, any subdivision of the meter will carry forward its dimensional imperfections.

The Micro Measurement System (MMS), on the other hand, approaches metrology from the other (microcosmic) direction. MMS' fundamental unit of measure is not the relatively crude meter, but a universal constant: the diameter of the protium nucleus with the proton at zero velocity, one unit each of MMS pressure (MMP 1) and Micro Measurement Gravity (MMG 1), and one degree above the temperature in which the cryogenic liquid state of protium is possible, as measured in Micro Measurement Temperature

[1] It was reported in 2010 that experiments led by Randolph Pohl at the Max Planck Institute for Quantum Optics had established the radius of a hydrogen proton at 0.84184 fm. (Doubling Pohl's number, of course, gives us the H proton's diameter.) Although this measurement has since been challenged by other scientists, it is among the closest meter-based values we have to date for the MMS constant.

(MMTe 1). (Note: the preceding MMS units, along with others based on various properties of protium, are explained in subsequent pages.)

The exact diameter of the protium nucleus (or proton) has not been established, although recent experiments at the Max Planck Institute suggest a value of 1.68378 femtometers (fm)[1]. Future science may be able to measure the protium nucleus with increasingly greater accuracy, but at the moment we can only make educated guesses about its diameter. So for now, the basic unit of measure in MMS is undefined in fractions of meter.

Inasmuch as MMS breaks ground in metrical areas long considered sacred by many scientists, we ask that you suspend any doubt you might have while considering this proposition with an open mind. I will close with what seems an appropriate observation from Lord Kelvin: *"If you cannot measure it, you cannot improve it."*

Thank you for your time and interest.

Josiah James Ingalls
January 24 2013
Austin, TX, USA

2 Grand Unified Theory Is Not Possible Without Unified Measurement

Grand unified theory is not possible without unified measurement. Before I give you my argument of why this is so, let me start by quoting the founder of IBM, Thomas J. Watson Sr.: "I can only see a market for maybe five computers worldwide." As many people know, Mr. Watson was one of those who built the first computer. He was considered a very knowledgeable person but as we all can see, there are far more computers in this world than he predicted. This brings me to my point that just because it is hard for society to imagine science and technology advancing to a level beyond what we can achieve now and what is currently necessary, doesn't mean it cannot and will not happen.

You cannot have a unified theory with no margins of error and no anomalies *unless* you have a unified measurement system. Traditionally, "unified measurement system" has meant one set of measurement systems used by all humanity; however, this is not what I am proposing. I propose one set of measurement systems based on a single, universal constant that can be accepted by the scientific community to empirically define something's existence. Otherwise, we are left with today's imperfect measurement systems that prevent us from developing and proving a grand unified theory in science.

Our current systems of measurement are based on different things that have certain characteristics that the scientific community considers acceptable as the basis of a given measurement system. Therefore, the unintentional consequence is that when you use more than one measurement system to scientifically define something's existence, you compile margins of error so that the more accurately you try to scientifically measure something, the less likely you are to accomplish it.

This dichotomy is reflected in the Heisenburg Principle in quantum mechanics. Heisenberg states there is a fundamental limit on the accuracy of measuring certain pairs of physical properties of a particle, such as position and momentum, so that both cannot be simultaneously known. In layman's terms, the more precisely one property is measured, the less precisely the other can be controlled, determined, or known. Using current systems based on separate things, you will always encounter this limitation. However, a unified measurement system where all measurements are based on a single thing with characteristics and properties that the scientific community deems acceptable for each type of measurement could virtually or actually limit uncertainty.

What laws of science would a unified measurement system have to obey? Since this seems to be new ground, I propose the following:

1. Justification for allowing a margin of error in a unified measurement system is not acceptable.
2. Organize the measurement system to scientifically define something's existence, the different types of measurements within the measurement system must be harmonious.
3. Systems of measurement shall be defined by a universal constant that does not add any margin of error to the measurement system.
4. In structuring a measurement system, accurate scientific data shall be used as the basis..
5. A measurement system's accuracy shall be verifiable by anyone using suitable equipment.
6. Have it be mandatory that each type of measurement be infinitely scalable, both up and down.

If a unified measurement system existed it would enable, for the first time, the tools necessary to disprove the uncertainty factor. The uncertainty factor is one of two places in science that still leaves room for a higher power. Some people in the religious community argue that since the Big Bang and everything else in the universe and in nature cannot be 100% scientifically defined, then a higher power must exist. My argument is that the universe has not been scientifically defined because of the lack of effort to create a unified measurement system, which some religious people may consider a challenge. However, I consider myself to be a scientist and firmly believe that if a higher power exists, it will be provable if we are able to scientifically define everything.

Most measurement formulas used today only work on earth because they are based on what we observe or what exists on our planet. For example, consider gravity and pressure. If I were working with a scientist on Mars on a scientific project that involved these two measurements, his gravitational pull and his atmospheric pressure (in megabars) would be different than mine. The current accepted formulas wouldn't work the same on Mars, which is a violation of law three above.

Other systems violate law six, because they were designed to go up but not down to infinity; therefore if you tried to use our current systems to measure the size of an atom of the hydrogen isotope protium, you could not get a correct measurement. The scientific and mathematics communities have generally expressed the viewpoint that it is not necessary or important to have a measurement system that is capable of measuring atoms by size, since measuring atoms by weight is sufficient.

If I have not yet convinced you that a newer and better measurement system is needed, consider this: The fact that our current measurement system cannot precisely measure the size of a hydrogen atom clearly shows that the system includes a margin of error. To find the source of this error, we must go back to the creation of our current English measurement system by King Henry I. Tradition holds that King Henry decreed that a yard should be the distance from the tip of his nose to the end of his outstretched thumb. Now, any person in today's society can clearly see that this measurement is not based on any scientific method whatsoever, and that our measurement of size should be defined with science.

The Metric System that succeeded King Henry's version is defined with far more scientific accuracy, but it too has a margin of error. The meter is now defined as the distance traveled by light in an absolute vacuum in $^1/_{299,792,458}$ of a second, which can never be truly accurate due to the margin of error in the measurement of time itself. For purposes of delineating the meter, the BIPM currently considers the HeNe laser wavelength to be as follows: $\lambda_{HeNe} = 632.99139822$ nm, with an estimated relative standard uncertainty *(U)* of 2.5×10^{-11}.

However, the speed of light is affected by temperature, and therefore is not a universal constant. This is another reason why the meter cannot be exactly defined by the distance that light travels in one second. This fact does not disprove Albert Einstein's $e=MC^2$, but it However, the speed of light is affected by temperature, and therefore is not a universal constant. This is another reason why the meter cannot be exactly defined by the distance that light travels in one second. This fact does not disprove Albert Einstein's $e=MC^2$, but it does demonstrate the scientific necessity of unifying temperature with space and time.

When you are using two or more systems together, a margin of error in one is repeated in all of the others. You may say that we are talking about quantities so small they do not matter, but theoretically that is like saying that the unit of measurement of inch number five on a ruler does not matter, that we should just skip from four to six. Clearly this is a ridiculous idea, which leads you to conclude that any quantity that exists in nature or science clearly matters, regardless of size.

This is why the concept of unified measurement systems is easy to accept, because it is in line with the very nature of quantities of any sort. Any person of rational thought, once the idea has been presented, can quickly come to the rational conclusion that this is a concept that should be widely adopted. The rational person then wonders why, if it makes such good sense, a unified measurement system has not been proposed before. It is my opinion that there are three primary reasons:

1. Children today are taught to accept every teaching as an absolute fact, and not question it.

2. Textbooks have been severely reduced in factual information to make them as simple as possible.

3. Historically, the people who defined our current measurements saw each system as unrelated to any others, since each of them measured different things in nature and the universe. Some denied science the advantages of unified measurement because they wanted their work to be recognized as unique (rather than an addition to someone else's work), and so devised idiosyncratic measurement systems.

It is irrelevant if some of today's technology is too inefficient to fully utilize all parts of the Micro Measurement Systems. In arguing why they should be widely adopted, I will cite an indisputable fact: the history of mathematics records many occasions in which someone's mathematical formulas were completely or partially adopted because they simplified the calculation of something, thus allowing them to be more used by general society instead of just those specializing in higher mathematics.

Let's use the electron volt as an example. One electron volt is equal to $1.602176565(35) \times 10^{-19}$ J under the current mathematical equation. In comparison, the Micro Measurement Electron volt System expresses this value simply as MMES 1. It is easy to see that the latter would be easier to comprehend by general society members not holding a degree in advanced mathematics.

The Micro Measurement Systems have been designed in a way that actually makes it suitable for international adoption, regardless of language. The language barrier has been the largest problem for the SI measurement systems due to the fact that its units are often given word names instead of being left as a solely mathematical formula. Therefore, as a result in translating from one language to the next, you end up with different meanings for the words. This is why in some countries the value of some of the SI units are not the same, so a gram in one country will be a different amount in another country. Another contributing factor to this anomaly is that since the word meaning varies by language, then the legal definition of the unit also varies from one country to the next, which creates a major problem in our modern day technological society. Society is traditionally used to having names for units of measure, but I caution the scientific community in trying to give names to the Micro Measurement Systems units of measure. No names were given to the units in the Micro Measurement Systems in order to prevent this problem from persisting. As a result, we end up with formulas that are solely mathematical. The only purpose of the initials in the formulas of measurement is to allow for an understanding of what is being calculated, especially given the fact that in simplifying the formulas in such a way that they calculate the same, it is important to have the identifying initials on all calculations to prevent confusion.

3 Proposed Revisions For the Formal Definitions of the Seven SI Base Units

The International Committee for Weights and Measures (CIMP) has proposed revisions for the formal definitions of the seven SI base units – second, metre, kilogram, ampere, kelvin, mole and candela. These revisions are being examined by CIPM and may be considered by the 25th General Conference on Weights and Measures in 2014. In keeping with the metric system tradition of deriving measurements from nature, the revisions are tied to a set of physical constants, e.g., the ground state of caesium 133 for the second; elementary electric charge for ampere; and Boltzmann and Avogadro constants for kelvin and mole, respectively.

The proposed revisions definitions of the seven S.I. base units and how they relate to one another.

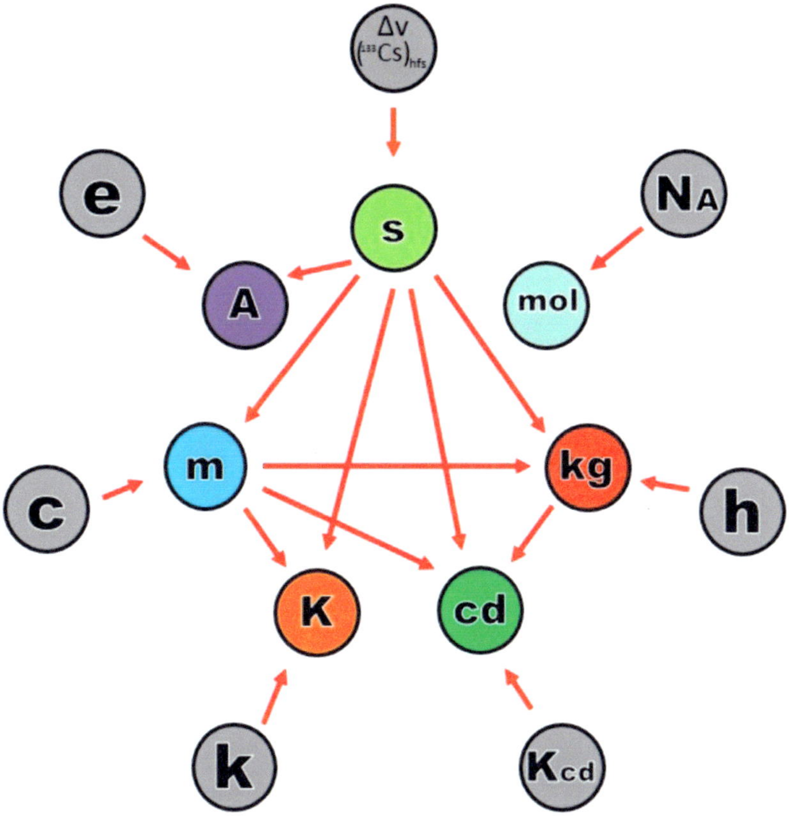

The proposed definitions of the Micro Measurement Systems base units and how they relate to one another.

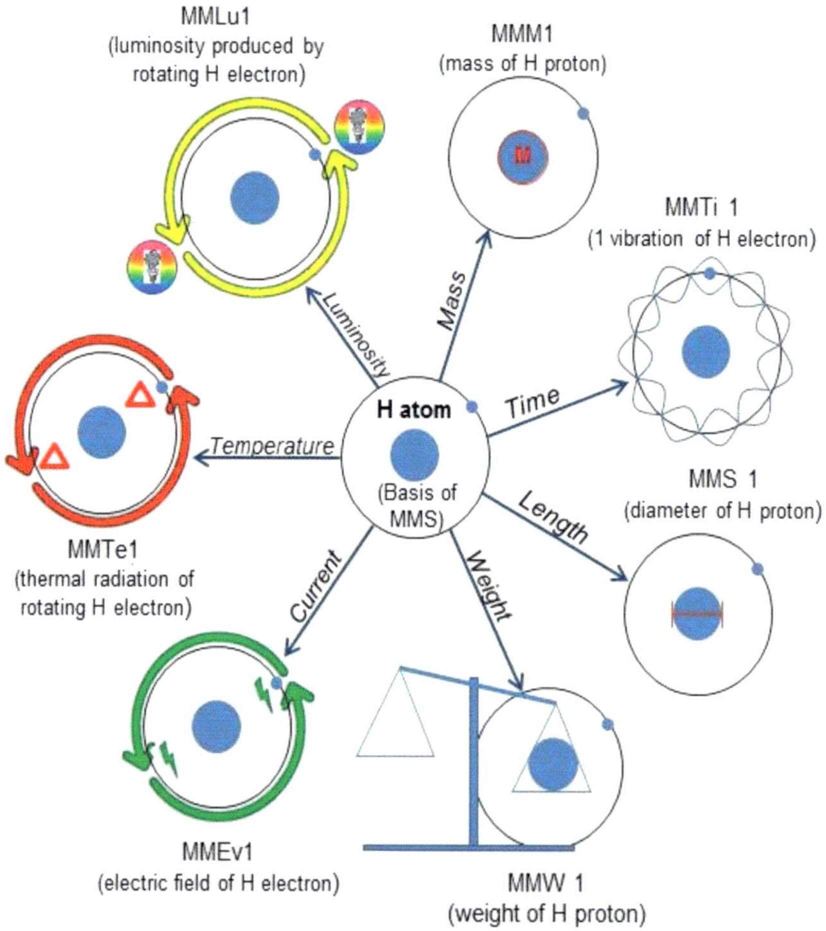

Base Unit	***Second***
Current Definition	The second is the duration of 9192631770 periods of the radiation corresponding to the transition between the two hyperfine levels of the ground state of the caesium-133 atom.
CIPM Proposed Definition	The second, s, is the unit of time; its magnitude is set by fixing the numerical value of the ground state hyperfine splitting frequency of the caesium-133 atom, at rest and at a temperature of 0 K, to be equal to exactly 9192631770 when it is expressed in the unit s^{-1}, which is equal to Hz.
MMS Unit	Micro Measurement Time System (MMTi 1)
MMS Unit Definition	MMTi 1 equals one vibration of the protium atom, measured at MMP 1, MMG 1, zero velocity, and 1 degree MMTe temperature above the point at which the cryogenic liquid state of protium is possible.
Metric Unit Comparison	One vibration of the protium atom (MMTi 1) equals 1 x 10^{-9} second.
MMS Unit	Because MMTi 1 is the smallest natural calculation of positive time, it is far

Advantage(s)	more precise. Plus, it doesn't require a human to interfere with the natural resting state of the atom, as is done with the atomic clock.
Base Unit	***Metre***
Current Definition	The metre is the length of the path travelled by light in vacuum during a time interval of 1/299792458 of a second.
CIPM Proposed Definition	The metre, m, is the unit of length; its magnitude is set by fixing the numerical value of the speed of light in vacuum to be equal to exactly 299792458 when it is expressed in the unit $m \cdot s^{-1}$.
MMS Unit	Micro Measurement Square (MMS 1)
MMS Unit Definition	MMS 1 is equal to the diameter[i] of the protium nucleus, at MMP 1 pressure, MMG 1 gravity, protium proton at zero velocity, and 1 degree MMte temperature above the point at which the cryogenic liquid state of protium is possible
Metric Unit Comparison	One MMS unit of distance (MMS 1) approximates 1.68 femtometres (fm) [ii].
MMS Unit Advantage(s)	MMS 1 is the smallest amount of physical distance yet presented as a metric, and it does not multiply the

	margin of error of time into its calculation of dimension.
Base Unit	***Kilogram***
Current Definition	The kilogram is the unit of mass; it is equal to the mass of the international prototype of the kilogram.
CIPM Proposed Definition	The kilogram, kg, is the unit of mass; its magnitude is set by fixing the numerical value of the Planck constant to be equal to exactly $6.62606X\times10^{-34}$ when it is expressed in the unit $s^{-1}\bullet m^{2}\bullet kg$, which is equal to J•s.
MMS Unit	Micro Measurement Weight System (MMW 1)
MMS Unit Definition	MMW 1 equals the weight of one protium nucleus (i.e., H proton), defined at MMG 1 gravity, proton at

[2] Based on work performed at Catalan Institute of Nanotechnology in Barcelona, Spain, and reported in 2012.. See http://www.dvice.com/archives/2012/04/how-do-you-weig.php.

	zero velocity, and one degree above the temperature in which the cryogenic liquid state of protium is possible using MMTe.
Metric Unit Comparison	The weight of a xenon proton is currently estimated as 1.7 yoctograms.[2] .
MMS Unit Advantage(s)	MMW 1 retains a standard of physical mass that is based on a physical constant that is not a relationship between two fields.
Base Unit	***Ampere***
Current Definition	The ampere is that constant current which, if maintained in two straight parallel conductors of infinite length, of negligible circular cross-section, and placed 1 m apart in vacuum, would produce between these conductors a force equal to 2×10^{-7} newton per metre of length.
CIPM Proposed Definition	The ampere, A, is the unit of electric current; its magnitude is set by fixing the numerical value of the elementary charge to be equal to exactly $1.60217X\times10^{-19}$ when it is expressed in the unit A•s, which is equal to C.
MMS Unit	Micro Measurement Electron volt System(MMEv 1)

MMS Unit Definition	MMEv 1 is the amount of electric field produced by one protium atom during one rotation of the nucleus. Parameters are defined at MMG 1, MMP 1, proton at zero velocity, and MMTe 1.
Metric Unit Comparison	Comparison remains to be developed.
MMS Unit Advantage(s)	MMEv 1 does not have the margin of error implicit in multiplying current times distance, and does not postulate an infinite condition.

Base Unit	***Kelvin***
Current Definition	The kelvin, unit of thermodynamic temperature, is the fraction 1/273.16 of the thermodynamic temperature of the triple point of water.
CIPM Proposed Definition	The kelvin, K, is the unit of thermodynamic temperature; its magnitude is set by fixing the numerical value of the Boltzmann constant to be equal to exactly $1.38065X\times10^{-23}$ when it is expressed in the unit $s^{-2}\cdot m^{2}\cdot kg\cdot K^{-1}$, which is equal to $J\cdot K^{-1}$.
MMS Unit	Micro Measurement Temperature System (MMTe 1)

MMS Unit Definition	One degree of MMTe (MMTe 1) equals the as the amount of thermal radiation produced by the rotation of an electron around the protium nucleus within MMTi 1. Parameters are defined at MMG 1, MMP 1, and proton at zero velocity.
Metric Unit Comparison	Comparison remains to be developed.
MMS Unit Advantage(s)	As the lowest quantum of positive temperature, MMTe 1 also offers a negative temperature system which allows for calculating temperatures below the theoretical absolute zero – a necessity since it is now known that molecular motion can continue below absolute zero under certain conditions. The metric system currently does not offer any proposal to address the issue of sub-absolute zero temperature.
Base Unit	***Mole***
Current Definition	The mole is the amount of substance of a system that contains as many elementary entities as there are atoms in 0.012 kilogram of carbon-12. When the mole is used, the elementary entities must be specified and may be atoms, molecules, ions, electrons, other particles, or specified groups of such

	particles.
CIPM Proposed Definition	The mole, mol, is the unit of amount of substance of a specified elementary entity, which may be an atom, molecule, ion, electron, any other particle or a specified group of such particles; its magnitude is set by fixing the numerical value of the Avogadro constant to be equal to exactly $6.02214X \times 10^{23}$ when it is expressed in the unit mol^{-1}.
MMS Unit	Micro Measurement Mass System (MMM 1)
MMS Unit Definition	MMM 1 is designed to measure mass in any configuration, including the mole. It is based on the mass of one protium nucleus. Parameters are set at MMG 1, protium proton at zero velocity, at MMTe 1.
Metric Unit Comparison	A meaningful comparison has yet to be developed. However, the mass of a proton has been estimated at 1.672 621 637(83) x 10^{-27} kg, or 938.272013(23) MeV/c^2, or 1.007 276 466 77(10) u (unified atomic mass units).[3]

[3] http://www.universetoday.com/57415/proton-mass/

MMS Unit Advantage(s)	The mole is useful in quantifying masses of like substances, as in moles of oxygen or tritium, but not as practical in discerning the mass of a disparate substance, such as a rock composed of many types of atoms, molecules and mixtures. By contrast, MMM 1 enables a precise calculation of a heterogeneous material mass through the use of the common denominator of the mass of a single proton – while also allowing exact measurements of each specific substance comprising that overall mass.
Base Unit	***Candela***
Current Definition	The candela is the luminous intensity, in a given direction, of a source that emits monochromatic radiation of frequency 540×10^{12} Hz and that has a radiant intensity in that direction of 1/**683 watt** per steradian.
CIPM Proposed Definition	The candela, cd, is the unit of luminous intensity in a given direction; its magnitude is set by fixing the numerical value of the luminous efficacy of monochromatic radiation of frequency 540×10^{12} Hz to be equal to exactly 683 when it is expressed in the

	unit s^3•m^{-2}•kg^{-1}•cd•sr, or cd•sr•W^{-1}, which is equal to lm•W^{-1}.
MMS Unit	MMLu 1
MMS Unit Definition	Micro Measurement Luminosity system measures all forms of luminous radiation waves. It is based on the amount of luminous radiation produced from a protium atom in the time it takes an electron to rotate around the nucleus once. Parameters are set at MMG 1, at MMP 1, proton at zero velocity, within MMTi 1, and at one degree above the temperature in which the cryogenic liquid state of protium is possible using MMTe.
Metric Unit Comparison	Comparison remains to be developed.
MMS Unit Advantage(s)	MMLu 1 measures only the amount of luminosity immediately produced during one period of electron rotation, irrespective of the time that the luminous radiation subsequently spends in transit or the distance it travels. This removes from the metric the margins of error implicit in calculations of time and distance. Also, MMLu 1 is the base unit for measuring luminosity from all sources at all intensities across the electromagnetic spectrum.

	Overall Note on MMS Units Each MMS unit carries two additional advantages: 1) Because any given MMS base unit is derived from a property of the same object (i.e., the protium atom), they do not multiply the margins of error that inevitably occur when using more than one formula to define metrics for physical or energetic phenomena. 2) Based as they are on the same object, MMS measurement units serve as checks upon each other and more precisely define one another.

The following illustration might be useful for appreciating the interrelationships among the seven MMS units described above. As shown, each unit is measured in turn, or checked, by two or more "sister" metrics that are also tied to singular properties of the hydrogen atom. This provides internal consistency for the entire system while minimizing potential margins for error.

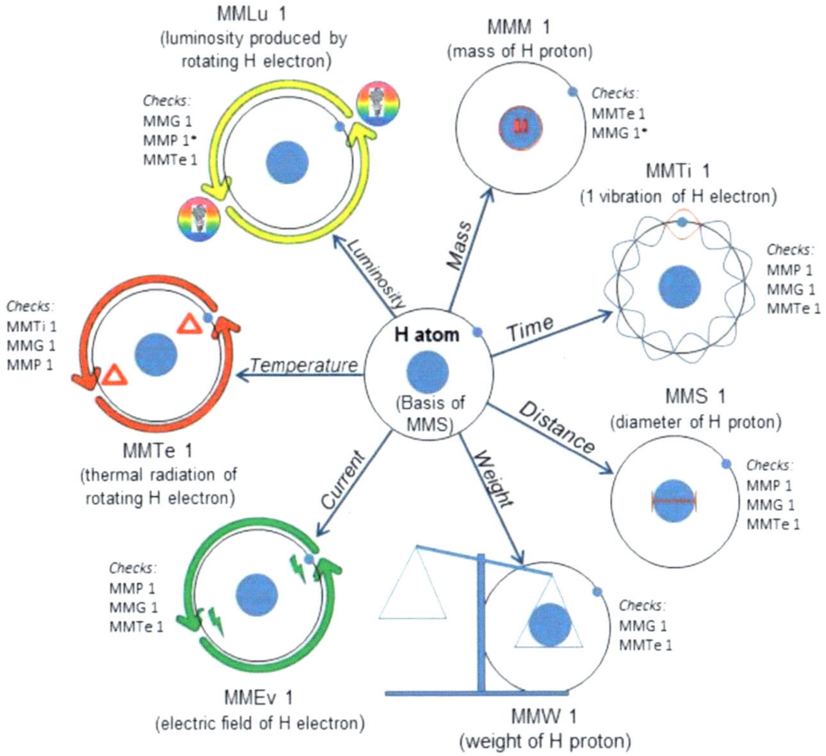

* MMP 1 and MMG 1, the MMS System metrics for pressure and gravity (respectively), are not included in this illustration for the sake of simplicity , and because they do not relate directly to the seven SI base units under discussion.

4 Advantages and Applications of MMS

As shown above, the protium atom is small and universally versatile enough to be considered the point of reference for MMS. No other measurement system has ever been designed like MMS. Throughout history, measurement systems have started at a human scale (such as a cubit, foot, or meter) and then have been calibrated up or down according to need. But never before has an extensible measurement system been based on a single universal constant, in this case the protium atom and its properties.

The micro measurement system is unique in that it does not use the decimal system used in base 10 mathematics. Instead, MMS uses as its base the number 14,400,000,000,000,000 (or 14.4×10^{14} in traditional scientific notation). The first column in a base 10 number can contain 1 to 9 units; add 1 digit, and the number is written as 10, with the new column representing 10s of units and the right-hand column still denoting individual units. Similarly MMS Level 1 (MMS 1) can represent 1 to 13,399,999,999,999,999 units. Add one more MMS unit to that total and the number flips to MMS 2 (with 14,400,000,000,000,000 units). While seemingly unwieldy to the inexperienced user, MMS 2 expresses an extremely large number in a very compact form that does not require exponential notation. (The exception is the calculation for distance, which uses a cubic notational system.) Also, it is important to remember that your base10 calculator software will not allow you to calculate MMS properly due to its reliance on decimals.

The metric system is destined to fail the needs of the scientific community, especially in quantum physics. The metric system defines its base units as phenomena that are independent of each other. However, we are quickly learning in quantum physics that nothing is independent and everything is interdependent. Yet even in the face of rising evidence of interdependence, the metric system establishment and of the world of metrology refuse to acknowledge this fundamental error in their world view. In my opinion, they are like the 19th century astronomers who continued to look for the "luminiferous ether" even after numerous experiments showed it was neither detectable nor necessary to explain celestial phenomena. It's time to stop trying to make the scientific data fit the ancient model and theory in metrology, and embrace a model of metrology that is in line with the scientific data.

5 Calculating Physical Objects

An advantage of MMS, in addition to its extreme precision, is its ability to calculate physical objects. When doing MMS calculations, it helps to keep a mental image of yourself counting physical objects instead of just compiling numbers; I find it helpful to imagine myself doing various calculations of grains of sand in my daughter's sandbox.

The inability of the current mathematical system to account precisely for physical objects is a fundamental problem. Let me give a simple example: I divide an apple into thirds, give one each to two friends and keep one for myself. Base 10 math tells me that we each now own 0.33 of the apple, which means that together we own 0.99 of an apple – but where did the .01 apple go? This problem persists even if I express one-third as 0.33 with a thousand more 3s behind it; there will always be a minute piece of apple that is not accounted for. While this inaccuracy may not matter much in the world of fruit lovers, it matters increasingly in quantum physics and other areas requiring extreme precision. A decimal approximation that calculates to infinity simply will not provide the precision called for in emerging science.

This brings me to a very important point. People assume that the number 3, or 100, or 101 for that matter, has value. But in fact these characters are nothing more than language, the same as A B or C; the numbers have no physical value unless you assign them to something with a physical value. Historically, languages have evolved because the existing language failed to express certain ideas. The fact that base 10 mathematics will not allow you to calculate physical objects exactly is evidence that it is inefficient as a language. So in order to overcome this hurdle, MMS does not use the decimal system.

6 Resolving *pi* And Defining the Point

Having no decimals, the micro measurement system offers far more accuracy. For example, for the first time since Pharoah's Egyptians tried to define *pi*, MMS offers a way to precisely calculate this ancient ratio. Also, it lets you define how many points are in a circle, allowing you to quantitatively subdivide the circle.

It has been my mission to do away with the margins of error that exist in the current measurement systems, but I also have seen that the decimal system introduces a margin of error in the language being used with existing measurement systems. So in order to ensure that the measurement systems that I've created would not be handicapped by the decimal system, I created new forms of calculations to complement my systems.

Defining the point is a fundamental problem in geometry and by extension, mathematics. Euclid, who created classical geometry, was quoted as saying that the point was too small to see and too small to write. Subsequent scholars took this to mean that the point had no dimensions, and proceeded to construct a mathematics based on the kind of ideal world that God was thought to live in. But in fact, you cannot have a system whose basic units have no dimensions and use it in a world of dimensions. Consider also that Euclid never said the point was without dimension or mass. In my opinion, he simply understood that the technology of his day was not capable of determining the point's metrics.

However, science has made great strides since Euclid and we are now able to scientifically define the point.

7 A System for Fractals

There are many mathematical benefits to defining the point as a precise quantum. One, it better enables you to mathematically calculate fractals when combined with the micro measurement system of distance.

The metric system's inability to handle and mathematically calculate fractals is a fundamental mathematical problem. This problem is solved once you define the point as MMS 1, combined with the very nature of how the micro measurement system's distance calculation is designed to be a fractal system. At first glance, you may not notice, but it is squares. Building on squares making larger squares and squares that are smaller than the previous squares, is strikingly similar to the definition of a fractal. [iii] So now through MMS, we finally have a way to begin to calculate fractals.

More precisely, however, there is another fundamental point of reference that really has never been defined in science. When you begin to get into advanced math, you realize that every measurement system has a point of reference, such as the point in time when you begin a journey, or the point at which you have calculable mass in the creation of the cosmos.

These points are assumed, but it is time to scientifically define them. And even if mass or distance or time has a smaller quantum than the defined point, it is irrelevant because the micro measurement system continues downward infinitely and continues upward infinitely.

8 Micro Measurement Square System

Micro Measurement Square system is a measurement system to measure distance. A single Micro Measurement Square unit is a precise square with each side being equal to the width of the nucleus of the hydrogen isotope protium, which shall be known as MMS 1 defined at MMP (Micro Measurement Pressure) 1, and at MMG (Micro Measurement Gravity) 1, with the proton at zero velocity, and at one degree above the temperature in which the cryogenic liquid state of protium is possible using MMTe (Micro Measurement Temperature). The next size up from MMS 1 shall be MMS 2 ,which is four Micro Measurement Squares forming a new square of the new size MMS 2 to the block of 2. The next level up consists of nine MMS 1s forming a new square and a new size, MMS 3, to the block of three.

To calculate the fourth block, you take the nine squares in MMS 3 and add to the right side four squares in a vertical line, and to the bottom row add three squares in a horizontal line, which will give you MMS 4, made up of 16 squares of MMS 1s. You can continue to calculate subsequent blocks to every size mathematically and scientifically possible.

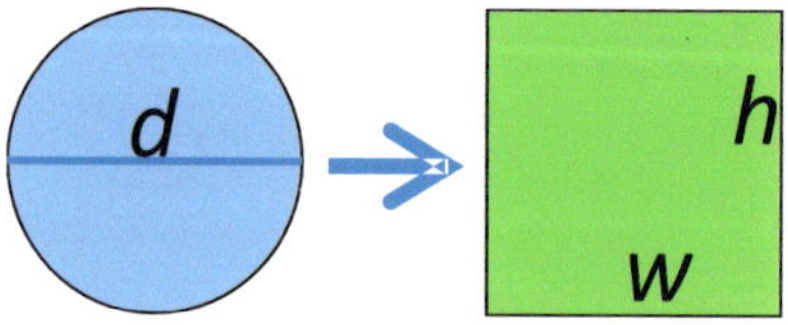

The diameter (*d*) of a protium nucleus defines the height (*h*) and width (*w*) of the Micro Measurement Square (MMS).

In the understanding that not every atom or molecule can be precisely measured in units of protium - because some are just micro-fractions bigger than a protium atom - we must go to a negative system. To get the next size down from MMS 1, you divide MMS 1 into 14,400,000,000,000,000 equal squares. Then subtract from the right side vertically 120,000,000 and from the bottom horizontally 119,999,999, which will leave you with a remainder of 14,399,999,760,000,001 (~14.399 x 10^{12}) squares which forms a new square and a new size which shall be called Micro Measurement Square One to the negative block of 119,999,999, expressed as MMS -1(119,999,999).

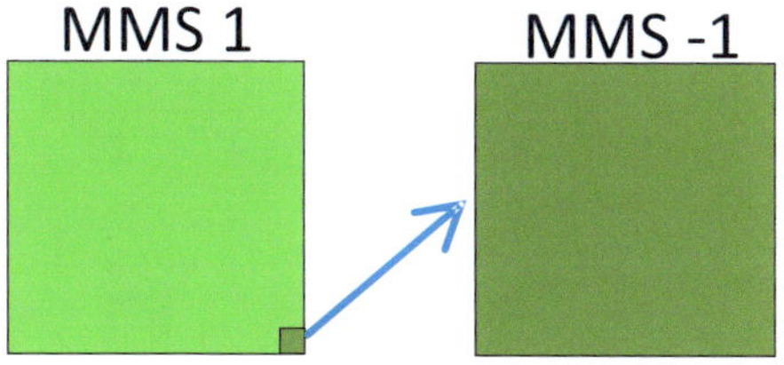

MMS 1 can be divided into 14.4 x 10^{14} equal squares.

Each of these squares = one unit of MMS -1.

When you get to MMS -1(1) you divide MMS -1(1) into 14,400,000,000,000,000 equal squares. Again, subtract from the right side vertically 120,000,000 and from the bottom horizontally 119,999,999 leaving a remainder of 14,399,999,760,000,001 ($\sim 14.399 \times 10^{12}$) squares which forms a new square and a new size which shall be called Micro Measurement Square 2 to the negative block of 119,999,999, or more succinctly MMS -2(119,999,999). This same method shall be repeated every time you get down to one single square, allowing for new measurements of size as far down as you would like to go.

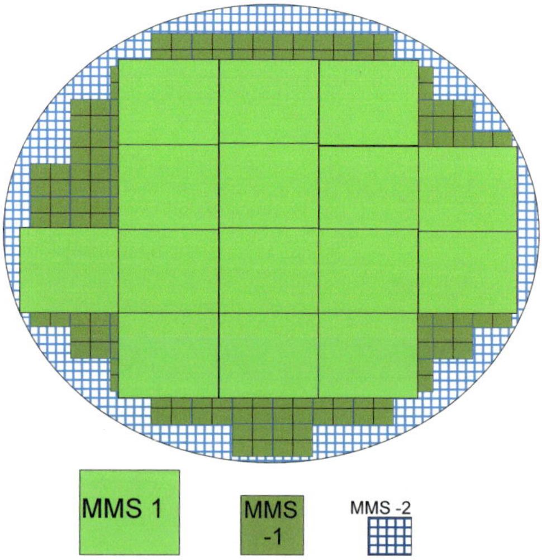

Note: MMS Squares Not to Scale

For circular objects, you can use MMS squares of different magnitudes to fill the area of the circle, and so measure it more precisely than by using our existing measurement systems.

Length and Width Measurements

To measure just the length or width of something using the formula above, you would add L or W to the end of your result. For example an MMS 1 size would now read MMS 1W or MMS 1L.

9 Cubed and Three Dimensional Measurements

This section is designed to assist in categorizing Micro Measurement Squares of any size in three dimensional (3D) objects. Out of a four-block square of Micro Measurement Squares the square block in the uppermost left hand corner shall be referred to as MMSB-1, which stands for Micro Measurement Square block one. The uppermost right hand corner shall be referred to as MMSB-2, which stands for Micro Measurement Square block two. The bottommost left hand corner shall be referred to as MMSB-3, which refers to Micro Measurement Square block three. The bottommost right hand corner shall be referred to as MMSB-4, representing Micro Measurement Square block four. In order to properly categorize your squares within any given size, number them left to right from the top line down to the bottom so the end result of an actual measurement would read MMSB-1 MMS 4.

Just add B after MMS for any measurement in order to properly categorize your squares within the measurement system.

To write out a calculation of a cubed square within the Micro Measurement Square System, use MMSC. So for example, an actual measurement would read MMSC 4; for length you would use MMSC 4l, and for width MMSC 4w. To properly categorize all of the existing squares within a cubed square, you would use layers.

Top layer shall be Layer 1, written as MMSCB L1; second layer would be MMSCB L2, and so on down the scale. You should number your blocks in each layer from left to right, starting with the uppermost left corner. So for example, an actual measurement would read MMSCB L2-3.

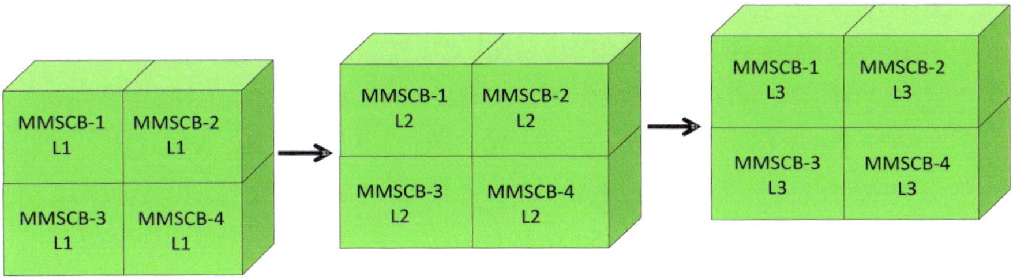

In the understanding that everything in nature and the universe is made up differently, it is important to allow each person the ability to express the final measurements in any way that they see fit, and not set rules within the measurement system, such as "always start with the largest size and then fill in the gaps," or "you should just start with the smallest size and then fill in the gaps." By *not* setting rules limiting interpretation, we allow for a more scientifically accurate measurement.

10 Micro Measurement Weight System

Micro Measurement Weight System According to some accounts, our current English system for weights is based on the weight of a grain seed plucked from the middle of an ear of barley. The old system seems not to have taken into account that all grains are not equal, and all are affected by moisture - leaving a margin of error that is huge by atomic standards. The Metric System defines weight using the gram, originally one millionth of the mass of a cubic meter of water, and currently defined as one thousandth of the mass of a specific object that is kept in a vault in France. However, there are efforts underway to redefine the gram in terms of physical quantities that could be reproduced in any laboratory with suitable equipment. Therefore the scientific community acknowledges a margin of error even in the venerable gram.

Obviously, the protium nucleus is not affected by moisture retention, nor does it need to be kept in a vault. Its weight is constant, therefore doing away with the margin of error in the measurement of weight. Accordingly, the Micro Measurement Weight (MMW) system is designed to measure weight without error. It is based on the weight of one protium nucleus, expressed as MMW 1 and defined at MMG 1, with the proton at zero velocity, and one degree above the temperature in which the cryogenic liquid state of protium is possible using MMTe.

The next size up from MMW 1 shall be MMW 2, consisting of 14,400,000,000,000,000 (14 x 10^{14}) units of MMW 1. MMW 3 shall consist of 28,800,000,000,000,000 (28.8 x 10^{14}) MMW 1s. Each MMW unit above MMW3 shall consist of 14,400,000,000,000,000 units more. Any remainder up to 14,399,999,999,999,999 shall be placed in parenthesis.

The first unit down from MMW 1 is MMW -1(14,399,999,999,999,999). When you get to MMW -1(1) you divide MMW -1(1) into 14,400,000,000,000,000 (14.4 x 10^{14}) equal parts. This will give you the next unit of weight down from MMW -1(1), which shall be MMW 2(14,399,999,999,999,999). This same method shall be repeated every time you get down to one, allowing for new measurements of weight as far down as you would like to go. For example, a final measurement would read MMW 100(35) MMW -3(99).

11 Micro Measurement Time System

Since Micro Measurement Square has done away with the margin of error in size, it eliminates one of the factors contributing to the error margin in measurement of time. This will allow measurement equipment to measure the vibrations of an atom more accurately. Given that in 2008 Albert Einstein's theory of $E=mc^2$ was declared to be true and accurate science, including the fact that the speed of time varies according to physical height, we must find another way to measure time.

Time is measured currently by the vibration of the cesium atom. That measure was selected in 1967, when Ce was considered to be the smallest stable atom. I argue that our technology wasn't advanced enough then to consider protium as stable. Since the measurement of time is actually a measurement of energy, the cesium atom can no longer be used, due to the fact that it is not the smallest measurement of energy according to $E = MC^2$. To properly measure time, you must track it with the smallest measurement of energy that still has the properties for unified measurement. The protium nucleus meets that requirement when defined at MMP 1, at MMG 1, with the proton at zero velocity, and at one degree above the temperature in which the cryogenic liquid state of protium is possible using MMTe. To properly write out Micro Measurement Time System you would use the abbreviation MMTi, then a space, then the number of vibrations of the protium nucleus.

The next scientific question to be asked is, at what point do you start counting the vibrations of protium? Your first thought would be to begin the counting of time from the first existence of mass within the galaxies, but you would quickly learn that it is impossible to determine at what point the first mass came to exist without unraveling the mystery of, and scientifically defining, the Big Bang.

If the measurement of time from the beginning is scientifically unattainable, it leads us to seek a more basic understanding of time itself. The idea of time was first accepted as a way to keep track of day and night, months and years, so to say that time is only the measurement of energy is false. In fact, true time is measured with energy to measure day, night, months and years, or the interval from one event to another. This brings us to the realization that true and correct time is in fact a parallel measurement of time, a measurement from one event to another, since each planet and or mass exists within parallel measurements of time.

For the purpose of creating a scientifically accurate measurement system that everyday society and the scientific community science can use and accept, one second shall be MMTi 11 or 10,000,000,000 (10^9) vibrations of the protium atom.[iv]

Accordingly,

MMTi 10 = 1,000,000,000 vibrations

MMTi 9 = 100,000,000 vibrations

MMTi 8 = 10,000,000 vibrations

MMTi 7 = 1,000,000, vibrations

MMTi 6 = 100,000, vibrations

MMTi 5 = 10,000 vibrations

MMTi 4 = 1,000 vibrations

MMTi 3 = 100 vibrations

MMTi 2 = 10 vibrations

MMTi 1 = 1 vibration

In order to measure amounts of time below a single vibration of the protium atom, you would use the Micro Measurement Square Negative System in order to calculate percentages of the distance which occurs in a single vibration. Although it is contrary to common practice to switch formulas, it is scientifically accurate to do so because we are no longer counting complete vibrations, only the percentage of the movement, which is distance. Thus, we must calculate by using the formula designed to measure distance. If we were to continue to measure down in the current formula of time, then you are implying that we are not counting distance, which in fact we are. The ability to scientifically define while being scientifically accurate will improve the science of time. For example, a measurement of time in MMS would read MMTi 30 MMS - 2(119,999,999).

12 Micro Measurement Temperature System

The current measurements of temperature are Celsius, Fahrenheit, Kelvin, and Rankine, each of which has a margin of error. To do away with this error, we will base the new system on the protium atom.

Micro Measurement Temperature system is a measurement system to measure thermal radiation. This system shall be defined as the amount of thermal radiation produced by the rotation of the electron around the protium nucleus, within the amount of time of MMTi 1 . This shall be the amount of MMTe calculated as MMTe 1 defined at MMG 1, at MMP 1, with the proton at zero velocity. The next size up from MMTe 1 shall be MMTe 2 which shall consist of 14,400,000,000,000,000 MMTe 1s. MMTe 3 shall consist of 28,800,000,000,000,000 MMTe 1s. Each MMTe unit up shall consist of 14,400,000,000,000,000 units more. The remainder up to 14,399,999,999,999,999 shall be placed in parenthesis. The next unit down from MMTe 1 is MMTe – 1(14,399,999,999,999,999). When you get to MMTe - 1(1) you divide MMTe -1(1) into 14,400,000,000,000,000 equally. This will give you the next unit of thermal radiation down from MMTe -1(1) which shall be MMTe -2(14,399,999,999,999,999). This same method shall be repeated every time you get down to one, allowing for new measurements of thermal radiation as far down as you would like to go. For example, a final measurement would read MMTe 100(35) MMTe -3(99).

13 Micro Measurement Pressure System

Micro Measurement Pressure System The current system of pressure measurement is primarily based on atmospheric pressure at sea level. This is not a universal constant, which is what we need to measure pressure exactly. The MMS pressure system shall be based on MMS 1, which is the amount of pressure exerted upon impact by one atom of protium traveling at the same speed at which the electron rotates around the nucleus, as defined at MMG 1, with the proton at zero velocity, and at one degree above the temperature in which the cryogenic liquid state of protium is possible using MMTe.

The next size up from MMP 1 shall be MMP 2, which shall consist of 14.400,000,000,000,000 (14.4×10^{12}) MMP 1s. MMP 3 shall consist of 28,800,000,000,000,000 MMP 1s. Each MMP unit up shall consist of 14,400,000,000,000,000 units more. The remainder up to 14,399,999,999,999,999 shall be placed in parenthesis.

The next unit down from MMPS 1 is MMP -1(14,399,999,999,999,999). When you get to MMP -1(1) you divide it into 14,400,000,000,000,000 equal parts. This will give you the next unit of pressure down from MMP -1(1) which shall be MMP -2(14,399,999,999,999,999). This same method shall be repeated every time you get down to one unit of scale, allowing for new measurements of pressure as far down as you would like to go. For example, a final measurement would read MMP 100(35) MMP -3(99).

14 Micro Measurement Liquid System

The Micro Measurement Liquid system is a measurement system to measure is defined as the amount of liquid of one atom protium atom in its cryogenic liquid state. This shall be the amount of liquid calculated as MMLi 1 as defined at MMG 1, with the proton at zero velocity, and at one degree below the temperature in which the cryogenic liquid state of protium is possible using MMTe. The next size up from MMLi 1 shall be MMLi 2 which shall consist of 14,400,000,000,000,000 (14.4×10^{14}) MMLi 1s. MMLi 3 shall consist of 28,800,000,000,000,000 (28.8×10^{14}) MMLi 1s. Each MMLi level up shall consist of 14.400,000,000,000,000 units more MMLi 1s. The remainder up to 14,399,999,999,999,999 shall be placed in parenthesis.
The next unit down from MMLi 1 is MMLi -1(14,399,999,999,999,999). When you get to MMLi -1(1) you divide MMLi -1(1) into 14,400,000,000,000,000 equally. This will give you the next unit of liquid down from MMLi -1(1) which shall be MMLi -- 2(14,399,999,999,999,999). This same method shall be repeated every time you get down to one, allowing for new measurements of liquid as far down as you would like to go. For example, a final measurement would read MMLi 100(35) MMLi -3(99).

15 Micro Measurement Gravity System

The Micro Measurement Gravity system shall be based on the amount of gravity produced by one protium atom between its nucleus and electron in the time it takes the electron to complete one rotation around the nucleus. This amount of gravity shall be calculated as MMG 1 as defined at MMP 1, with the proton at zero velocity, and at one degree above the temperature in which the cryogenic liquid state of protium is possible using MMTe.

The next size up from MMG 1 shall be MMG 2, which shall consist of 14,400,000,000,000,000 (14.4 x 10^{14}) MMG 1s. MMG 3 shall consist of 28,800,000,000,000,000 MMG 1s. Each MMG level up shall consist of 14,400,000,000,000,000 units more. The remainder up to 14.399,999,999,999,999 shall be placed in parenthesis.

The next unit down from MMG 1 is MMG -1(13.399,999,999,999,999). When you get to MMG -1(1) you divide MMG -1(1) into 14,400,000,000,000,000 equally. This will give you the next unit of gravity down from MMG -1(1), which shall be MMG -2(14,399,999,999,999,999). This same method shall be repeated every time you get down to one, allowing for new measurements of gravity as far down as you would like to go. For example, a final measurement would read MMG 100(35) MMG -3(99).

With the discovery of electromagnetism, the scientific

community abandoned the idea of measuring electric field and magnetic field as separate calculations. However, the unification of the electric field and magnetic field into a single calculation known as electromagnetism has produced problems in measuring high energy frequency and low energy frequency in the same system of units. There is also the problem created by the discovery that a magnetic field can exist without an electric field. This is why it is important to consider the electric field and magnetic field in physics as separate properties that work together to create an electromagnetic field. To solve the calculation problems, it is necessary to have three separate measurement formulas for electric field, magnetic field and the electromagnetic field.

16 Micro Measurement Electron Volt System

Micro Measurement Electron volt system (MMEv) is designed to measure electric field. This system shall be defined as the amount of electric field produced by one protium atom in the amount of time it takes the electron to complete one rotation of the nucleus. This shall be the amount of electric field calculated as MMEv 1 defined at MMG 1, at MMP 1, with the proton at zero velocity, within MMTi 1, and at one degree above the temperature in which the cryogenic liquid state of protium is possible using MMTe.

The next size up from MMEv 1 shall be MMEv 2, which shall consist of 14,400,000,000,000,000 (14.4 x 10^{14}) MMEv 1s. MMEv 3 shall consist of 28,800,000,000,000,000 (28.8 x 10^{14}) MMEv 1s. Each MMEv unit up shall consist of 14.400,000,000,000,000 units more. The remainder up to 14.399,999,999,999,999 shall be placed in parenthesis.

The next unit down from MMEv 1 is MMEv -1(14,399,999,999,999,999). When you get to MMEv -1(1) you divide MMEv -1(1) into 14,400,000,000,000,000 equally. This will give you the next unit of electric field down from MMEv -1(1) which shall be MMEv -2(14,399,999,999,999,999). This same method shall be repeated every time you get down to one, allowing for new measurements of electric field as far down as you would like to go. For example, a final measurement would read MMEv 100(35) MMEv -3(99).

17 Micro Measurement Mass System

Micro Measurement Mass system is a designed to measure mass. This system shall be defined as the amount of mass of one protium nucleus. This shall be the amount of Mass calculated as MMM 1 defined at MMG 1, with the proton at zero velocity, and at one degree above the temperature in which the cryogenic liquid state of protium is possible using MMTe.

The next size up from MMM 1 shall be MMM 2, which shall consist of 14,400,000,000,000,000 (14.4×10^{14}) MMM 1s. MMM 3 shall consist of 28,800,000,000,000,000 MMM 1s ((28.8×10^{14}) Each MMM unit up shall consist of 14,400,000,000,000,000 units more. The remainder up to 14,399,999,999,999,999 shall be placed in parenthesis.

The next unit down from MMM 1 is MMM -1(14,399,999,999,999,999). When you get to MMM -1(1) you divide MMM -1(1) into 14,400,000,000,000,000 equally. This will give you the next unit of mass down from MMM -1(1) which shall be MMM -2(14,399,999,999,999,999). This same method shall be repeated every time you get down to one, allowing for new measurements of mass as far down as you would like to go. For example, a final measurement would read MMM 100(35) MMM -3(99).

18 Micro Measurement Magnetic Field System

Micro Measurement Magnetic field system is designed to measure magnetic field. This system shall be defined as the amount of magnetic field produced by one protium atom in the amount of time it takes the electron to complete one rotation of the nucleus. This shall be the amount of magnetic field calculated as MMMf 1 defined at MMG 1, at MMP 1, with the proton at zero velocity, within MMTi 1, and at one degree above the temperature in which the cryogenic liquid state of Protium is possible using MMTe.

The next size up from MMEf 1 shall be MMMf 2, which shall consist of 14,400,000,000,000,000 (14.4 x 10^{14}) MMMf 1s. MMMf 3 shall consist of 28,800,000,000,000,000 (28.8 x 10^{14}) MMEf 1s. Each MMEf unit up shall consist of 14.400,000,000,000,000 units more. The remainder up to 14.399,999,999,999,999 shall be placed in parenthesis.

The next unit down from MMMf 1 is MMMf -1(14,399,999,999,999,999). When you get to MMEf -1(1) you divide MMMf -1(1) into 14,400,000,000,000,000 equally. This will give you the next unit of magnetic field down from MMMf -1(1) which shall be MMMf -2(14,399,999,999,999,999). This same method shall be repeated every time you get down to one, allowing for new measurements of magnetic field as far down as you would like to go. For example, a final measurement would read MMMf 100(35) MMMf -3(99).

19 Micro Measurement Electromagnetic Field System

Micro Measurement Electromagnetic field system designed to measure electromagnetic field. This system shall be defined as the amount of electric field and magnetic field combined of equal amounts, calculated as electromagnetic field produced by one protium atom in the amount of time it takes the electron to complete one rotation of the nucleus. This shall be the amount of electromagnetic field calculated as MMEf 1 defined at MMG 1, at MMP 1, with the proton at zero velocity, within MMTi 1, and at one degree above the temperature in which the cryogenic liquid state of protium is possible using MMTe.

The next size up from MMEf 1 shall be MMEf 2, which shall consist of 14,400,000,000,000,000 (14.4×10^{14}) MMEf 1s. MMEf 3 shall consist of 28,800,000,000,000,000 (28.8×10^{14}) MMEf 1s. Each MMEf unit up shall consist of 14.400,000,000,000,000 units more. The remainder up to 14.399,999,999,999,999 shall be placed in parenthesis.

The next unit down from MMEf 1 is MMEf -1(14,399,999,999,999,999). When you get to MMEf -1(1) you divide MMEf -1(1) into 14,400,000,000,000,000 equally. This will give you the next unit of electromagnetic field down from MMEf -1(1) which shall be MMEf -2(14,399,999,999,999,999). This same method shall be repeated every time you get down to one, allowing for new measurements of electromagnetic field as far down as you would like to go. For example, a final measurement would read MMEf 100(35) MMEf -3(99).

20 Micro Measurement Speed System

The speed of light is an amount of light which humanity has decided is important. It currently is measured by the distance in which light travels from its source to a specific point. However it is necessary to create a formula for speed based on a universal constant of speed and distance found in nature instead of a percentage of distance which science has predetermined. Such a formula allows you to measure speed in a way that is more congruent with everything else found in nature. The hydrogen atom is a perfect example of a precise measurement of time and a precise measurement of distance traveled. If you measure speed by the distance in which the electron rotates around the nucleus once with the amount of time it takes to do so, then both elements of speed are based on a single universal constant.

Micro Measurement Speed system is a measurement system to measure speed. This system shall be defined as the amount of distance in which the electron rotates around the nucleus once in the amount of time it takes to do so using the protium atom. This shall be the amount of MMSp calculated as MMSp 1 defined at MMG 1, at MMP 1,with the proton at zero velocity, and at one degree above the temperature in which the cryogenic liquid state of protium is possible using MMTe. The next size up from MMSp 1 shall be MMSp 2 which shall consist of 14,400,000,000,000,000 MMSp 1s. MMSp 3 shall consist of 28,800,000,000,000,000 MMSp 1s. Each MMSp unit up shall consist of 14,400,000,000,000,000 units more. The remainder up to 14,399,999,999,999,999 shall be placed in parenthesis. The next unit down from MMSp 1 is MMSp -1(14,399,999,999,999,999). When you get to MMSp -1(1) you divide MMSp -1(1) into 14,400,000,000,000,000 equally. This will give you the next unit of speed down from MMSp -1(1) which shall be MMSp -2(14,399,999,999,999,999). This same method shall be repeated every time you get down to one, allowing for new measurements of speed as far down as you would like to go. For example, a final measurement would read MMSp 100(35) MMSp -3(99).

21 Micro Measurement Luminosity System

Light from hydrogen atom, which is one kind of electromagnetic radiation caused by moving charged particles, is the most important data for understanding the structure of the atom.

Micro Measurement Luminosity system is a measurement system to measure all forms of Luminous radiation waves. This system shall be defined as the amount of luminous radiation produced from a protium atom in the amount of time it takes electron to rotates around the nucleus once. This shall be the amount of MMLu calculated as MMLu 1 defined at MMG 1, at MMP 1, at with the proton at zero velocity, within MMTi 1, and at one degree above the temperature in which the cryogenic liquid state of protium is possible using MMTe. The next size up from MMLu 1 shall be MMLu 2 which shall consist of 14,400,000,000,000,000 MMLu 1s. MMLu 3 shall consist of 28,800,000,000,000,000 MMLu 1s. Each MMLu unit up shall consist of 14,400,000,000,000,000 units more. The remainder up to 14,399,999,999,999,999 shall be placed in parenthesis. The next unit down from MMLu 1 is MMLu -1(14,399,999,999,999,999). When you get to MMLu -1(1) you divide MMLu -1(1) into 14,400,000,000,000,000 equally. This will give you the next unit of Luminous radiation down from MMLu -1(1) which shall be MMLu -2(14,399,999,999,999,999). This same method shall be repeated every time you get down to one, allowing for new measurements of luminous radiation as far down as you would like to go. For example, a final measurement would read MMLu 100(35) MMLu -3(99). If more than one wavelength is present (as is usually the case), one must sum or integrate over the spectrum of wavelengths present to get the total luminous intensity.

22 Parallel Measurements

Using the methods of measurement described in the above primary formulas, you can take any conceivable size or amount of a quantity and measure up or down with that, which gives a new formula of measurement. For example, if you take MMS 120,000,000 which is a two dimensional square, and you desire to measure using that measurement to the block of four, the final results would read Micro Measurement Square to the block of 120,000,000 Parallel Measurement to the block of four, which shall be known as MMS 120,000,000 PM 4. If you are going to use a negative size it would read Micro Measurement Square Two to the block of negative 119,999,999 Parallel Measurement to the block of four, which shall be known as MMS - 2(119,999,999) PM 4. For this example we have used Micro Measurement Square; however, when using other formulas you would follow the guidelines in writing out those formulas, just adding the PM numbered calculation in the appropriate place.

With the completion of all fourteen measurement systems, we have done away with a cumulative margin of error that has not been considered by the general scientific community. Each of these systems relies on one or more related systems in order to develop an exact unit of measure. Without this uniformity, you would create margins of error that would diminish your ability to be accurate.

23 Defense of the Theory

Before I give you my defense of this theory, let me start by quoting Sir Isaac Newton:

"Do not create anything new unless you are prepared to defend it for all time."

We will proceed with a question-and-answer format.

Q. *The MMS formulas are designed to subdivide infinitely; but isn't it scientifically impossible to subdivide a physical object infinitely?*

A. On a subatomic level that is a correct statement. However, to scientifically define something's existence, your measurement system must be able to subdivide past the point where it is physically impossible for an object to be divided any further. Why? Because future measurement technology is likely to make that point a moving target. Therefore, the measurement system must be able to subdivide infinitely.

Q. *How do you calculate for human error in measurement systems designed to have no margins of error?*

A. The Micro Measurement Systems were designed for future technology to measure for humans, and not for humans to measure by hand.

Q. *How do you measure precisely and accurately when the technology being used to measure is not capable of* *measuring accurately on a molecular level?*

A. We use current technology to produce more advanced equipment that enables new generations of technology to emerge. This self-reinforcing process eventually will lead us to a point where the tools used for measurement will be more accurate on a molecular level. This is another reason why the MMS formulas are designed for future technology.

Q. *Say you are trying to make a metal rod with a single row of atoms 45 billion atoms wide. You could not get a scientifically perfect measurement for that metal rod because if most of the electrons are to the right, then it would be shorter on the left side. So how do you measure precisely to make the rod?*

A. You cannot make the rod precisely the width of the specified number of atoms. You can measure the rod precisely using the formulas, but due to the nature of atoms the rod will never be precisely and constantly the same size. Each time you measure the rod you will come up with a different measurement – unless all of the electrons around each nucleus are in the exact same positions each time you measure. That is nearly impossible today, but future technology may increase the possibility.

[2]***Q.*** *Are you saying that if we scientifically define everything through absolute measurement, that which cannot be measured demonstrates the existence of a higher power?*

A. No. What I am saying is that if we have the ability to measure everything in infinite space with no future technology needed, once we have measured everything in infinite space with no margins of error and no anomalies, humans will have the capability to prove that a higher power exists; but before we can even begin to get close we need a set of measurement systems designed to have no anomalies and no margins of error that are unified around one universal constant in order to scientifically define everything.

Q. *If temperature was unified with space and time, would this allow for the meter to be accurately measured by the distance in which light travels in* $^{1}/_{299,792,458}$ *of a second?*

A. The speed of light can never be an absolute constant. If you unify temperature and the effects of thermal radiation waves with space and time, you must now also take into consideration the environment around the light wave, regardless of where it is in its travels out from the original source of light. This means you are taking into consideration the surrounding atoms and molecules, but since all molecules are made up of atoms then we are actually just talking about atoms. Since we have established that thermal radiation waves affect the speed of light, then you must also take into consideration what produces thermal radiation waves, which are also atoms. This now brings us to the conclusion that we need to look at what causes the atoms to produce thermal radiation waves. It is the gravitational pull that the nucleus has on the electron which moves the electron around the nucleus, but the rate at which it moves is affected by the atom's mass. The electron now in movement around the nucleus produces electric field. This electric field now produces electromagnetic field, which is now also producing light itself and thermal energy, but most of these fields and the amounts that are being produced by the atom, are affected by pressure. In order to truly measure the speed of light, it must be placed in context with space-time-gravity-mass-electric field-magnetic field-temperature-pressure. Now we have unified all of these formulas in order to calculate the speed of light. Even if the light wave never actually comes in contact with a single atom, you must recalculate its speed with every atom that it passes and the different amounts of the different elements that we are now calculating that exist around the light wave. Einstein unified space and time, and here we have now unified all of these phenomena to measure the speed of light, which proposes a new question in itself: Why didn't Einstein propose it in his theory of space and time? In my opinion, he failed to see the connections between $E=mc^2$ and his work on light and how light travels in its full potential. It is logical to conclude that it is possible for him not to have seen it even though it was right in front of him, and for this conjecture I give an example. The thought experiment known as "Einstein's clock in a box" was disproven by his own formula $E=mc^2$ because he simply forgot to calculate $E=mc^2$ for the clock into the equation.

24 Conclusion

In summary, MMS may offer a precise new way to measure the increasingly extreme dimensions we're encountering (for example, in semiconductor science) by utilizing the interrelated properties of a single object– the H proton – which itself is infinitesimal and universally constant. The system is a little ahead of its time in that some of its units cannot yet be measured – but at least two units (for length and weight) can already be expressed in conventional metrics and it may not be long before the others are known as well.

MMS challenges many basic assumptions, which can be dangerous ground. However, I believe that too much of our science is based on assumptions instead of fact or data. That is, a scientist makes an assumption and builds an entire theory around it, adding supporting data and mathematics as they arise to give credence to the idea. However, the assumption itself is never questioned. An example is the Euclidean point, which has never been defined but which nonetheless has given rise to an entire branch of mathematics. In this respect, Euclidean geometry might be considered a form of "phantom science."

Phantom science has a long and distinguished history, filled with expansive assumptions supported by clever arguments that ultimately failed for lack of scientific evidence. One is reminded of the chemists who argued convincingly that a substance called phlogiston caused things to burn, until proof of oxidation set their world aflame. Likewise were the geocentric astronomers whose increasingly complex apologies of planetary motions broke down under heliocentric analysis.

But even today phantom science, held aloft by expansive postulates based on scant evidence, is still with us. We cannot explain why the cosmos seems to be missing large amounts of mass, so we conjecture the existence of dark matter and dark energy. We don't understand the apparent weakness of gravity compared to other basic forces, so we invent string theory. Time will tell if either of these theories have substance, but we may ultimately find that both suffer from the same weakness: the effects of cumulative margins of error in our measurement systems.

It is here that I think the next great frontier of science lies: in the development of a uniform, self-corrective measurement system that does away with margins of error, and with them the apparent anomalies that generate so many fanciful theories. A single unified measurement theory may hold the greatest potential for solving the perplexing anomalies that today are causing many scientists to indulge in quasi-scientific concepts about unobservable phenomena that resemble exercises in metaphysical thought. They remind me of the poet Hughes Mearns' description of an elusive entity:

> "Yesterday, upon the stair,
> I met a man who wasn't there…
> When I came home last night at three,
> The man was waiting there for me

> But when I looked around the hall,
> I couldn't see him there at all!"

Perhaps we would do better to remember Lord Kelvin's admonishment that measurement is the true substance of science, and that without reliable measurement we scientists are not much more than philosophers and storytellers. As Lord Kelvin himself put it:

> "In physical science, a first essential step in the direction of learning any subject is to find principles of numerical reckoning and practicable methods for measuring some quality connected with it. I often say that when you can measure what you are speaking about, and express it in numbers, you know something about it; but when you cannot measure it, when you cannot express it in numbers, your knowledge is of a meagre and unsatisfactory kind; it may be the beginning of knowledge, but you have scarcely in your thoughts advanced to the stage of *science*, whatever the matter may be."

In the spirit of Lord Kelvin, MMS offers us the opportunity to put precise measurement once again at the center of scientific achievement. MMS represents a paradigm shift in science because it is does not have room for fantasy or argument from authority. Indeed, MMS is designed to instill a mindset that causes a person to question an assertion no matter what the authority of the person who makes it, and to insist instead on seeing the data that supports it.

In my opinion, educated people cannot call themselves scientists if they accept the assumptions that their area is based on, and never ask themselves what happens when the assumptions are not true, or what they can do to prove or disprove those assumptions. Indeed, this would be a paradigm shift. I dream that we, as scientists, will embrace this new way of thinking; and when we do, the scientific community as a whole will begin to make breakthroughs in areas that have eluded us for centuries.

May Science be redefined!

wwwibmms.org

25 About the Author

Josiah Ingalls is a devoted physics researcher and inventor who has had a lifetime fascination with the science of measurement. His Micro Measurement System is the culmination of more than seven years of research and development aimed at crafting an ultimate metrology capable of extremely precise calculations extensible to infinity. Born in Bangor, Maine, Ingalls grew up in central Texas, where he worked in several fields before starting his own company. He and his family reside in Austin, TX, where he is an active member of Austin Physics Group, Loyal Order of Moose, Toastmasters, and Gay and Lesbian Chamber of Commerce.

[i] It was reported in 2010 that experiments led by Randolph Pohl at the Max Planck Institute for Quantum Optics had established the radius of a hydrogen proton at 0.84184 fm. (Doubling Pohl's number, of course, gives us the H proton's diameter to be 1.68368 fm.) Although this measurement has since been challenged by other scientists, it is among the closest meter-based values we have to date for the MMS constant.

[iii] The Merriam-Webster online dictionary defines fractal as "any of various extremely irregular curves or shapes for which any suitably chosen part is similar in shape to a given larger or smaller part when magnified or reduced to the same size."

Some factual information in this document was quoted and or used as reference material from Wikipedia as well as Webster's dictionary, as well as information release by the BIPM.

Made in the USA
Monee, IL
04 May 2026

49457544R00048